Beginners Guide to Liquid Chromatography

Joseph C. Arsenault, Patrick D. McDonald, Ph.D.

Waters

THE SCIENCE OF WHAT'S POSSIBLE.™

Acknowledgments

We wish to thank our colleagues: Dr. Uwe Neue, Dr. Diane Diehl, David Collis, Dr. Mark Baynham, Geoffrey McConnell, Micah Watt, and Allysa Waldropt for their critical reading of the manuscript and many helpful suggestions; Melissa Clark, Susan Corman, Ian Hanslope, and Victoria Walton for their illustrations and layout; and, Natalie Crosier for editing the book and coordinating its production.

Table of Contents

Note: Additional information on words in red throughout text can be found in this section.

List of Figures

List of Tables

What Is Liquid Chromatography?
Brief History and Definition

Liquid chromatography was defined in the early 1900's by the work of the Russian botanist, Mikhail S. Tswett. His pioneering studies focused on separating compounds [leaf pigments], extracted from plants using a solvent, in a column packed with particles.

Tswett filled an open glass column with particles. Two specific materials that he found useful were powdered chalk [calcium carbonate] and alumina. He poured his sample [solvent extract of homogenized plant leaves] into the column and allowed it to pass into the particle bed. This was followed by pure solvent. As the sample passed down through the column by gravity, different colored bands could be seen separating because some components were moving faster than others. He related these separated, different-colored bands to the different compounds that were originally contained in the sample. He had created an analytical separation of these compounds based on the differing strength of each compound's chemical attraction to the particles. The compounds that were more strongly attracted to the particles *slowed down*, while other compounds more strongly attracted to the solvent *moved faster*. This process can be described as follows: the compounds contained in the sample distribute, or partition differently between the moving solvent, called the mobile phase, and the particles, called the stationary phase. This causes each compound to move at a different speed, thus creating a separation of the compounds.

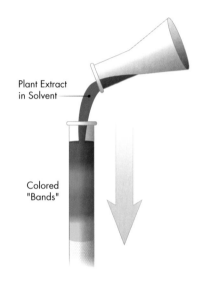

Plant Extract in Solvent

Colored "Bands"

Figure A: Tswett's Experiment

Tswett coined the name *chromatography* [from the Greek words *chroma*, meaning color, and *graph*, meaning writing—literally, *color writing*] to describe his colorful experiment. [Curiously, the Russian name Tswett means *color*.] Today, liquid chromatography, in its various forms, has become one of the most powerful tools in analytical chemistry.

Liquid Chromatography [LC] Techniques

Liquid chromatography can be performed using planar [Techniques 1 and 2] or column techniques [Technique 3]. Column liquid chromatography is the most powerful and has the highest capacity for sample. In all cases, the sample first must be dissolved in a liquid that is then transported either onto, or into, the chromatographic device.

Technique 1. The sample is spotted onto, and then flows through, a thin layer of chromatographic particles [stationary phase] fixed onto the surface of a glass plate [Figure B]. The bottom edge of the plate is placed in a solvent. Flow is created by capillary action as the solvent [mobile phase] diffuses into the dry particle layer and moves up the glass plate. This technique is called thin-layer chromatography or TLC.

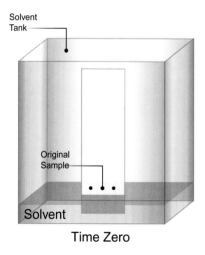
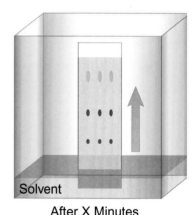

Figure B: Thin-Layer Chromatography

Note that the apparently *black* sample is a mixture of FD&C yellow, red and blue food dyes that has been chromatographically separated.

Technique 2. In Figure C, samples are spotted onto paper [stationary phase]. Solvent [mobile phase] is then added to the center of the spot to create an outward radial flow. This is a form of paper chromatography. [Classic paper chromatography is performed in a manner similar to that of TLC with linear flow.] In the upper image, the same, black, FD&C dye sample is applied to the paper.

Yellow 5

Blue 1 / Yellow 5

Red 40

Red 3

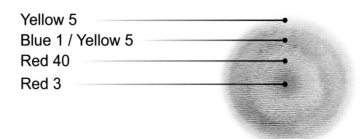

Paper chromatography reveals that a sample mixture that appears black is actually composed of several dyes.

Red 40

Blue 1

Further testing on a single pair shows that a 'purple' mixture can be separated into individual blue and red dyes.

Blue 1 / Yellow 5

Another experiment confirms that this paper system cannot separate the blue and yellow dyes; they appear as a single 'green' spot. Collectively, these tests prove that, except for the green ring, each colored circular band in the 'black' sample at the top represents a single dye.

Figure C: Paper Chromatography

Notice the difference in separation power for this particular paper system when compared to the TLC plate. The green ring indicates that the paper cannot separate the yellow and blue dyes from each other, but this separation is not a problem with the TLC system. Conversely, the paper LC system can separate the two red dyes, while in the chosen TLC system, they appear as a single spot. We will discuss in a later section how selectivity may be optimized to create desired separations in LC.

Technique 3. In this, the most powerful approach, the sample passes through a column or a disposable cartridge device containing appropriate particles [stationary phase]. These particles are called the chromatographic packing material. Solvent [mobile phase] flows through the device. In solid-phase extraction [SPE], the sample is loaded onto the cartridge and the solvent stream carries the sample through the device. As in Tswett's experiment, the compounds in the sample are then separated by traveling at different individual speeds through the device. Here the *black* sample is loaded onto a cartridge. Different solvents are used in each step to create the separation.

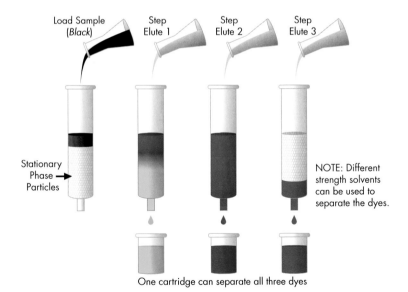

Figure D–1: Column Chromatography – Solid-Phase Extraction [SPE] . As shown here, SPE is often done on single-use, disposable plastic cartridge columns.

When the cartridge format is used, there are several ways to achieve flow. Gravity or vacuum can be used for columns that are not designed to withstand pressure. Typically, the particles in this case are larger in diameter [> 50 microns] so that there is less resistance to flow. Open glass columns [Tswett's experiment] are an example of this. In addition, small, plastic columns, typically in the shape of syringe barrels, can be filled with packing-material particles and used to perform sample preparation. This is called solid-phase extraction [SPE]. Here, the chromatographic device, called a cartridge, is used, usually with vacuum-assisted flow, to clean up a very complex sample before it is analyzed further.

Smaller particle sizes [<10 microns] are required to improve separation power. However, smaller particles have greater resistance to flow, so higher pressures are needed to create the desired solvent flow rate. Pumps and columns designed to withstand high pressure are necessary. When moderate to high pressure is used to flow the solvent through the chromatographic column, the technique is called HPLC.

Figure D–2: HPLC Column

What Is High-Performance Liquid Chromatography [HPLC]?

The acronym *HPLC*, coined by the late Prof. Csaba Horváth for his 1970 Pittcon® paper, originally indicated the fact that high pressure was used to generate the flow required for liquid chromatography in packed columns. In the beginning, pumps only had a pressure capability of 500 psi [35 bar]. This was called *high-pressure* liquid chromatography [HPLC]. The early 1970s saw a tremendous leap in technology. These new HPLC instruments could develop up to 6,000 psi [400 bar] of pressure, and incorporated improved injectors, detectors, and columns. HPLC really began to take hold in the mid-to late-1970s. With continued advances in performance during this time [smaller particles, even higher pressure], the acronym remained the same, but the name was changed to *high-performance* liquid chromatography [HPLC].

High-performance liquid chromatography [HPLC] is now one of the most powerful tools in analytical chemistry. It has the ability to separate, identify, and quantitate the compounds that are present in any sample that can be dissolved in a liquid. Today, compounds in trace concentrations as low as *parts per trillion* [ppt] may easily be identified. HPLC can be, and has been, applied to just about any sample, such as pharmaceuticals, food, nutraceuticals, cosmetics, environmental matrices, forensic samples, and industrial chemicals.

What Is Ultra-Performance Liquid Chromatography [UPLC® Technology]?

In 2004, further advances in instrumentation and column technology were made to achieve very significant increases in resolution, speed, and sensitivity in liquid chromatography. Columns with smaller particles [1.7 micron] and instrumentation with specialized capabilities designed to deliver mobile phase at 15,000 psi [1,000 bar] were needed to achieve a new level of performance. A new system had to be holistically created to perform ultra-performance liquid chromatography [UPLC® technology].

Basic research is being conducted today by scientists working with columns containing even smaller, 1-micron-diameter particles and instrumentation capable of performing at 100,000 psi [6,800 bar]. This provides a glimpse of what we may expect in the future.

How Does a High-Performance Liquid Chromatograph Work?
HPLC System Diagram

The components of a basic high-performance liquid chromatography [HPLC] system are shown in the simple diagram in Figure E.

A reservoir holds the solvent [called the mobile phase, because it moves]. A high-pressure pump [solvent delivery system or solvent manager] is used to generate and meter a specified flow rate of mobile phase, typically milliliters per minute. An injector [sample manager or autosampler] is able to introduce [inject] the sample into the continuously flowing mobile phase stream that carries the sample into the HPLC column. The column contains the chromatographic packing material needed to effect the separation. This packing material is called the stationary phase because it is held in place by the column hardware. A detector is needed to *see* the separated compound bands as they elute from the HPLC column [most compounds have no color, so we cannot see them with our eyes]. The mobile phase exits the detector and can be sent to waste, or collected, as desired. When the mobile phase contains a separated compound band, HPLC provides the ability to collect this fraction of the eluate containing that purified compound for further study. This is called preparative chromatography [discussed in the section on HPLC Scale].

Note that high-pressure tubing and fittings are used to interconnect the pump, injector, column, and detector components to form the conduit for the mobile phase, sample, and separated compound bands.

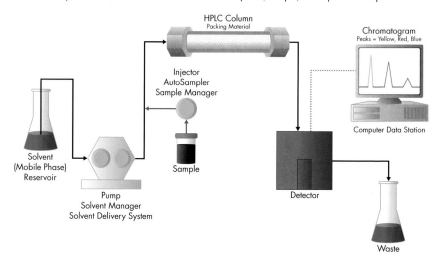

Figure E: High-Performance Liquid Chromatography [HPLC] System

The detector is wired to the computer data station, the HPLC system component that records the electrical signal needed to generate the chromatogram on its display and to identify and quantitate the concentration of the sample constituents (see Figure F). Since sample compound characteristics can be very different, several types of detectors have been developed. For example, if a compound can absorb ultraviolet light, a UV-absorbance detector is used. If the compound fluoresces, a fluorescence detector is used. If the compound does not have either of these characteristics, a more universal type of detector is used, such as an evaporative-light-scattering detector [ELSD]. The most powerful approach is the use multiple detectors in series. For example, a UV and/or ELSD detector may be used in combination with a mass spectrometer [MS] to analyze the results of the chromatographic separation. This provides, from a single injection, more comprehensive information about an analyte. The practice of coupling a mass spectrometer to an HPLC system is called LC/MS.*

For more information, see MS Primer at <www.waters.com/primers> . Also, see #6 on list of References for Further Reading on p. 40.

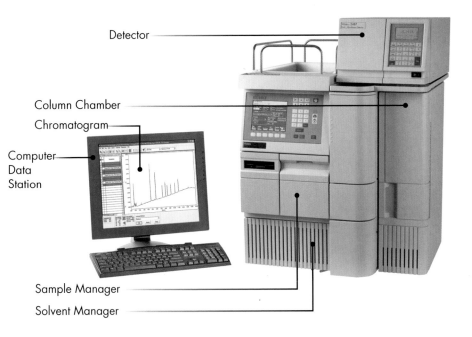

Detector

Column Chamber

Chromatogram

Computer
Data
Station

Sample Manager

Solvent Manager

Figure F: A Typical HPLC [Waters Alliance®] System.

HPLC Operation

A simple way to understand how we achieve the separation of the compounds contained in a sample is to view the diagram in Figure G.

Mobile phase enters the column from the left, passes through the particle bed, and exits at the right. Flow direction is represented by green arrows. First, consider the top image; it represents the column at time zero [the moment of injection], when the sample enters the column and begins to form a band. The sample shown here, a mixture of yellow, red, and blue dyes, appears at the inlet of the column as a single black band,

Injected Sample Band (blue, red & yellow mixture appears *black*)

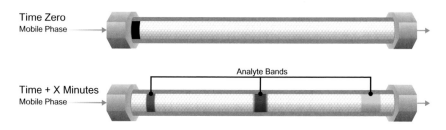

Figure G: Understanding How a Chromatographic Column Works – Bands

[In reality, this sample could be anything that can be dissolved in a solvent; typically the compounds would be colorless and the column wall opaque, so we would need a detector to see the separated compounds as they elute.]

After a few minutes [lower image], during which mobile phase flows continuously and steadily past the packing material particles, we can see that the individual dyes have moved in separate bands at different speeds. This is because there is a competition between the mobile phase and the stationary phase for attracting each of the dyes or analytes. Notice that the yellow dye band moves the fastest and is about to exit the column. The yellow dye likes [is attracted to] the mobile phase more than the other dyes. Therefore, it moves at a *faster* speed, closer to that of the mobile phase. The blue dye band likes the packing material more than the mobile phase. Its stronger attraction to the particles causes it to move significantly *slower*. In other words, it is the most retained compound in this sample mixture. The red dye band has an intermediate attraction for the mobile phase and therefore moves at an *intermediate* speed through the column. Since each dye band moves at different speed, we are able to separate it chromatographically.

What Is a Detector?

As the separated dye bands leave the column, they pass immediately into the detector. The detector contains a flow cell that *sees* [detects] each separated compound band against a background of mobile phase [see Figure H]. [In reality, solutions of many compounds at typical HPLC analytical concentrations are colorless.] An appropriate detector has the ability to sense the presence of a compound and send its corresponding electrical signal to a computer data station. A choice is made among many different types of detectors, depending upon the characteristics and concentrations of the compounds that need to be separated and analyzed, as discussed earlier.

What Is a Chromatogram?

A chromatogram is a representation of the separation that has chemically [chromatographically] occurred in the HPLC system. A series of peaks rising from a baseline is drawn on a time axis. Each peak represents the detector response for a different compound. The chromatogram is plotted by the computer data station [see Figure H].

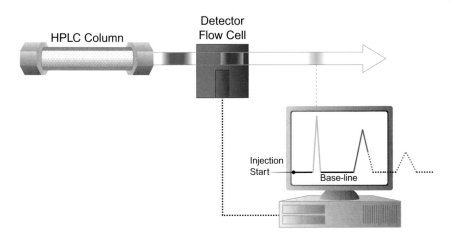

Figure H: How Peaks Are Created

In Figure H, the yellow band has completely passed through the detector flow cell; the electrical signal generated has been sent to the computer data station. The resulting chromatogram has begun to appear on screen. Note that the chromatogram begins when the sample was first injected and starts as a straight line set near the bottom of the screen. This is called the baseline; it represents pure mobile phase passing through the flow cell over time. As the yellow analyte band passes through the flow cell, a stronger signal is sent to the computer. The line curves, first upward, and then downward, in proportion to the concentration of the yellow dye in the sample band. This creates a peak in the chromatogram. After the yellow band passes completely out of the detector cell, the signal level returns to the baseline; the flow cell now has, once again, only pure mobile phase in it. Since the yellow band moves fastest, eluting first from the column, it is the first peak drawn.

A little while later, the red band reaches the flow cell. The signal rises up from the baseline as the red band first enters the cell, and the peak representing the red band begins to be drawn. In this diagram, the red band has not fully passed through the flow cell. The diagram shows what the red band and red peak would look like if we stopped the process at this moment. Since most of the red band has passed through the cell, most of the peak has been drawn, as shown by the solid line. If we could restart, the red band would completely pass through the flow cell and the red peak would be completed [dotted line]. The blue band, the most strongly retained, travels at the slowest rate and elutes after the red band. The dotted line shows you how the completed chromatogram would appear if we had let the run continue to its conclusion. It is interesting to note that the width of the blue peak will be the broadest because the width of the blue analyte band, while narrowest on the column, becomes the widest as it elutes from the column. This is because it moves more slowly through the chromatographic packing material bed and requires more time [and mobile phase volume] to be eluted completely. Since mobile phase is continuously flowing at a fixed rate, this means that the blue band widens and is more dilute. Since the detector responds in proportion to the concentration of the band, the blue peak is lower in height, but larger in width.

Identifying and Quantitating Compounds

In Figure H, three dye compounds are represented by three peaks separated in time in the chromatogram. Each elutes at a specific location, measured by the elapsed time between the moment of injection [time zero] and the time when the peak maximum elutes. By comparing each peak's retention time [t_R] with that of injected reference standards in the same chromatographic system [same mobile and stationary phase], a chromatographer may be able to identify each compound.

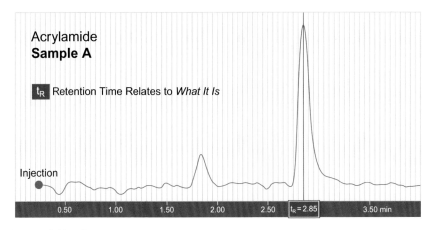

Acrylamide
Sample A

t_R Retention Time Relates to *What It Is*

Injection

0.50   1.00   1.50   2.00   2.50 $t_R = 2.85$ 3.50 min

Figure I–1: Identification

In the chromatogram shown in Figure I–1, the chromatographer knew that, under these LC system conditions, the analyte, acrylamide, would be separated and elute from the column at 2.85 minutes [retention time]. Whenever a new sample, which happened to contain acrylamide, was injected into the LC system under the same conditions, a peak would be present at 2.85 minutes [see Sample B in Figure I–2].

[For a better understanding of why some compounds move more slowly [are better retained] than others, please review the HPLC Separation Modes section on page 29].

Once identity is established, the next piece of important information is how much of each compound was present in the sample. The chromatogram and the related data from the detector help us calculate the concentration of each compound. The detector basically responds to the concentration of the compound band as it passes through the flow cell. The more concentrated it is, the stronger the signal; this is seen as a greater peak height above the baseline.

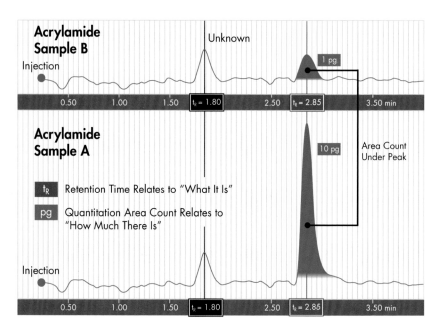

Figure I–2: Identification and Quantitation

In Figure I–2, chromatograms for Samples A and B, on the same time scale, are stacked one above the other. The same volume of sample was injected in both runs. Both chromatograms display a peak at a retention time [t_R] of 2.85 minutes, indicating that each sample contains acrylamide. However, Sample A displays a much bigger peak for acrylamide. The area under a peak [peak area count] is a measure of the concentration of the compound it represents. This area value is integrated and calculated automatically by the computer data station. In this example, the peak for acrylamide in Sample A has 10 times the area of that for Sample B. Using reference standards, it can be determined that Sample A contains 10 picograms of acrylamide, which is ten times the amount in Sample B [1 picogram]. Note there is another peak [not identified] that elutes at 1.8 minutes in both samples. Since the area counts for this peak in both samples are about the same, this unknown compound may have the same concentration in both samples.

Isocratic and Gradient LC System Operation

Two basic elution modes are used in HPLC. The first is called isocratic elution. In this mode, the mobile phase, either a pure solvent or a mixture, *remains the same throughout the run.* A typical system is outlined in Figure J–1.

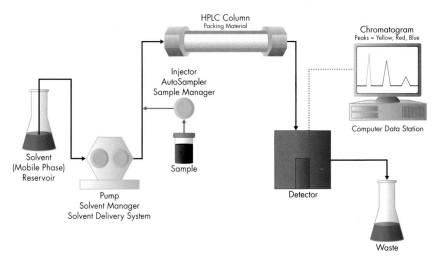

Figure J–1: Isocratic LC System

The second type is called gradient elution, wherein, as its name implies, *the mobile phase composition changes during the separation.* This mode is useful for samples that contain compounds that span a wide range of chromatographic polarity [see section on HPLC Separation Modes]. As the separation proceeds, the elution strength of the mobile phase is increased to elute the more strongly retained sample components.

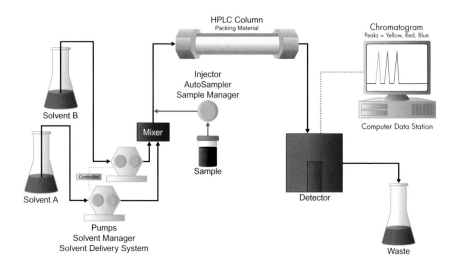

Figure J–2: High-Pressure-Gradient System

In the simplest case, shown in Figure J–2, there are two bottles of solvents and two pumps. The speed of each pump is managed by the gradient controller to deliver more or less of each solvent over the course of the separation. The two streams are combined in the mixer to create the actual mobile phase composition that is delivered to the column over time. At the beginning, the mobile phase contains a higher proportion of the weaker solvent [Solvent A]. Over time, the proportion of the stronger solvent [Solvent B] is increased, according to a predetermined timetable.

Note that in Figure J–2, the mixer is downstream of the pumps; thus the gradient is created under *high pressure*. Other HPLC systems are designed to mix multiple streams of solvents under *low pressure*, ahead of a single pump. A gradient proportioning valve selects from the four solvent bottles, changing the strength of the mobile phase over time, as shown in Figure J–3.

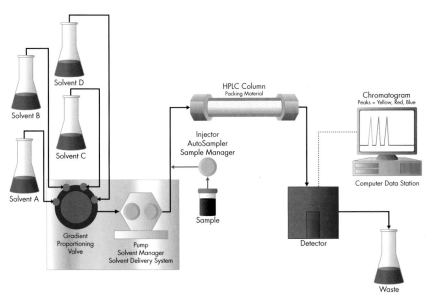

Figure J–3: Low-Pressure-Gradient System

HPLC Scale [Analytical, Preparative, and Process]

We have discussed how HPLC provides analytical data that can be used both to identify and to quantify compounds present in a sample. However, HPLC can also be used to purify and collect desired amounts of each compound, using a fraction collector downstream of the detector flow cell. This process is called preparative chromatography [see Figure K].

In preparative chromatography, the scientist is able to collect the individual analytes as they elute from the column [e.g., in this example: yellow, then red, then blue].

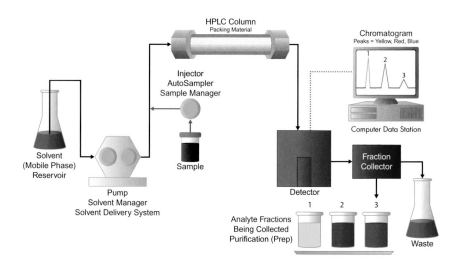

Figure K: HPLC System for Purification: Preparative Chromatography

The fraction collector selectively collects the eluate, that now contains a purified analyte, for a specified length of time. The vessels are moved so that each collects only a single analyte peak.

A scientist determines goals for purity level and amount. Coupled with knowledge of the complexity of the sample and the nature and concentration of the desired analytes relative to that of the matrix constituents, these goals, in turn, determine the amount of sample that needs to be processed and the required capacity of the HPLC system. In general, as the sample size increases, the size of the HPLC column will become larger and the pump will need higher volume-flow-rate capacity. Determining the capacity of an HPLC system is called selecting the HPLC *scale*. Table A lists various HPLC scales and their chromatographic objectives.

Scale	Chromatographic Objective
Analytical	Information [compound ID and concentration]
Semi–preparative	Data and a small amount of purified compound [< 0.5 gram]
Preparative	Larger amounts of purified compound [> 0.5 gram]
Process [Industrial]	Manufacturing quantities [grams to kilograms]

Table A: Chromatography Scale

The ability to maximize selectivity with a specific combination of HPLC stationary and mobile phases—achieving the largest possible separation between two sample components of interest—is critical in determining the requirements for scaling up a separation [see discussion on HPLC Separation Modes]. Capacity then becomes a matter of scaling the column volume [V_C] to the amount of sample to be injected and choosing an appropriate particle size [determines pressure and efficiency; see discussion of Separation Power]. Column volume, a function of bed length [L] and internal diameter [i.d.], determines the amount of packing material [particles] that can be contained (see Figure L).

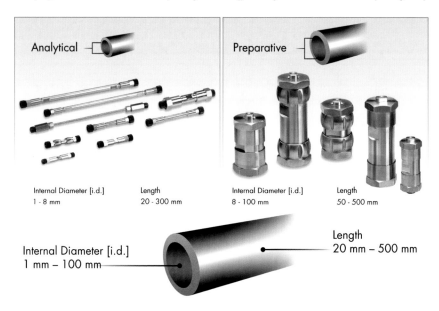

Analytical

Preparative

| Internal Diameter [i.d.] | Length | Internal Diameter [i.d.] | Length |
| 1 - 8 mm | 20 - 300 mm | 8 - 100 mm | 50 - 500 mm |

Internal Diameter [i.d.]
1 mm – 100 mm

Length
20 mm – 500 mm

Figure L: HPLC Column Dimensions

In general, HPLC columns range from 20 mm to 500 mm in length [L] and 1 mm to 100 mm in internal diameter [i.d.]. As the scale of chromatography increases, so do column dimensions, especially the cross-sectional area. To optimize throughput, mobile phase flow rates must increase in proportion to cross-sectional area. If a smaller particle size is desirable for more separation power, pumps must then be designed to sustain higher mobile-phase-volume flow rates at high backpressure. Table B presents some simple guidelines on selecting the column i.d. and particle size range recommended for each scale of chromatography.

For example, a semi-preparative-scale application [red X] would use a column with an internal diameter of 10–40 mm containing 5–15 micron particles. Column length could then be calculated based on how much purified compound needs to be processed during each run and on how much separation power is required.

Scale	Column Diameter				Particle Size microns
	1–8 mm	10–40 mm	50–100 mm	> 100 mm	
Analytical	X				1.7–10
Semi-Prep		X			5–15
Prep			X		15–100
Process				X	100+

Table B: Chromatography Scale vs. Column Diameter and Particle Size

A software-based Waters Prep Calculator CD is available to help you properly scale the size of your column and set other operating parameters, such as flow rate, as you scale up your separation. Visit <http:// www.waters.com/prepcalc>.

HPLC Column Hardware
Design

A column tube and fittings must contain the chromatographic packing material [stationary phase] that is used to effect a separation. It must withstand backpressure created both during manufacture and in use. Also, it must provide a well-controlled [leak-free, minimum-volume, and zero-dead-volume] flow path for the sample at its inlet, and analyte bands at its outlet, and be chemically inert relative to the separation system [sample, mobile, and stationary phases]. Most columns are constructed of stainless steel for highest pressure resistance. PEEK™ [an engineered plastic] and glass, while less pressure tolerant, may be used when inert surfaces are required for special chemical or biological applications. [Figure M–1].

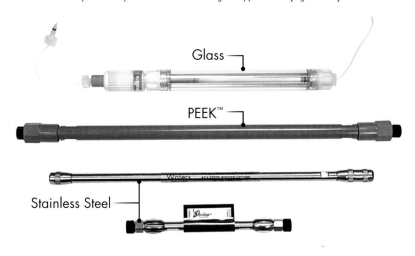

Figure M–1: Column Hardware Examples

A glass column wall offers a visual advantage. In the photo in Figure M–2, flow has been stopped while the sample bands are still in the column. You can see that the three dyes in the injected sample mixture have already separated in the bed; the yellow analyte, traveling fastest, is just about to exit the column.

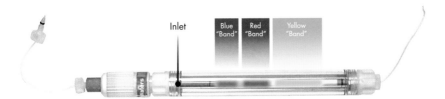

Figure M–2: A Look Inside a Column

Separation Performance – Resolution

The degree to which two compounds are separated is called chromatographic resolution [R_s]. Two principal factors that determine the overall separation power or resolution that can be achieved by an HPLC column are: mechanical separation power, created by the column length, particle size, and packed-bed uniformity, and chemical separation power, created by the physicochemical competition for compounds between the packing material and the mobile phase. Efficiency is a measure of mechanical separation power, while selectivity [α] is a measure of chemical separation power.

Mechanical Separation Power – Efficiency

If a column bed is stable and uniformly packed, its mechanical separation power is determined by the column length and the particle size. Mechanical separation power, also called efficiency, is often measured and compared by a plate number [symbol = N]. Smaller-particle chromatographic beds have higher efficiency and higher backpressure. For a given particle size, more mechanical separation power is gained by increasing column length. However, the trade-offs are longer chromatographic run times, greater solvent consumption, and higher backpressure. Shorter column lengths minimize all these variables but also reduce mechanical separation power, as shown in Figure N.

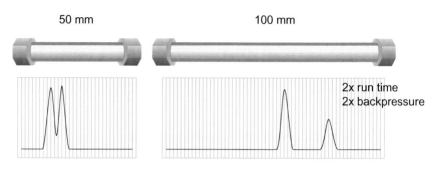

Figure N: Column Length and Mechanical Separating Power [Same Particle Size]

For a given particle chemistry, mobile phase, and flow rate, as shown in Figure O, a column of the same length and i.d., but with a smaller particle size, will deliver more mechanical separation power *in the same time*. However, its backpressure will be much higher.

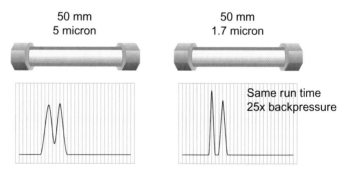

Figure O: Particle Size and Mechanical Separating Power [Same Column Length]

Chemical Separation Power – Selectivity

The choice of a combination of particle chemistry [stationary phase] and mobile-phase composition—the separation system—will determine the degree of chemical separation power [how we change the speed of each analyte]. Optimizing selectivity is the most powerful means of creating a separation; this may obviate the need for the brute force of the highest possible mechanical efficiency. To create a separation of any two specified compounds, a scientist may choose among a multiplicity of phase combinations [stationary phase and mobile phase] and retention mechanisms [modes of chromatography]. These are discussed in the next section.

HPLC Separation Modes

In general, three primary characteristics of chemical compounds can be used to create HPLC separations. They are:
- Polarity
- Electrical Charge
- Molecular Size

First, let's consider polarity and the two primary separation modes that exploit this characteristic: normal-phase and reversed-phase chromatography.

Separations Based on Polarity

A molecule's structure, activity, and physicochemical characteristics are determined by the arrangement of its constituent atoms and the bonds between them. Within a molecule, a specific arrangement of certain atoms that is responsible for special properties and predictable chemical reactions is called a functional group. This structure often determines whether the molecule is *polar* or *non-polar*. Organic molecules are sorted into classes according to the principal functional group(s) each contains. Using a separation mode based on polarity, the relative chromatographic retention of different kinds of molecules is largely determined by the nature and location of these functional groups. As shown in Figure P, classes of molecules can be ordered by their relative retention into a range or spectrum of chromatographic polarity from highly polar to highly non-polar.

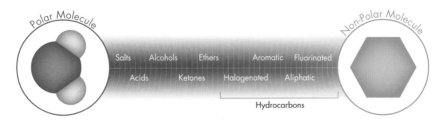

Figure P: Chromatographic Polarity Spectrum by Analyte Functional Group

Water [a small molecule with a high dipole moment] is a polar compound. Benzene [an aromatic hydrocarbon] is a non-polar compound. Molecules with similar chromatographic polarity tend to be attracted to each other; those with dissimilar polarity exhibit much weaker attraction, if any, and may even repel one another. This becomes the basis for chromatographic separation modes based on polarity.

Another way to think of this is by the familiar analogy: oil [non-polar] and water [polar] don't mix. Unlike in magnetism where opposite poles attract each other, chromatographic separations based on polarity depend upon the stronger attraction between likes—and the weaker attraction between opposites. Remember, *"Like attracts like"* in polarity-based chromatography.

Mobile Phase ⟶ Stationary Phase / Packing Material ⟶

Figure Q: Proper Combination of Mobile and Stationary Phases Effects Separation Based on Polarity

To design a chromatographic separation system [see Figure Q], we create competition for the various compounds contained in the sample by choosing a mobile phase and a stationary phase with different polarities. Then, compounds in the sample that are similar in polarity to the stationary phase [column packing material] will be delayed because they are more strongly attracted to the particles. Compounds whose polarity is similar to that of the mobile phase will be preferentially attracted to it and move faster.

In this way, based upon differences in the relative attraction of each compound for each phase, a separation is created by changing the speeds of the analytes.

Figures R–1, R–2, and R–3 display typical chromatographic polarity ranges for mobile phases, stationary phases, and sample analytes, respectively. Let's consider each in turn to see how a chromatographer chooses the appropriate phases to develop the attraction competition needed to achieve a polarity-based HPLC separation.

Figure R–1: Mobile Phase Chromatographic Polarity Spectrum

A scale, such as that shown in Figure R-1, upon which some common solvents are placed in order of relative chromatographic polarity is called an eluotropic series. Mobile phase molecules that compete effectively with analyte molecules for the attractive stationary phase sites displace these analytes, causing them to move faster through the column [weakly retained]. Water is at the polar end of mobile-phase-solvent scale, while hexane, an aliphatic hydrocarbon, is at the non-polar end. In between, single solvents, as well as mis-cible-solvent mixtures [blended in proportions appropriate to meet specific separation requirements], can be placed in order of elution strength. Which end of the scale represents the 'strongest' mobile phase depends upon the nature of the stationary phase surface where the competition for the analyte molecules occurs.

Figure R–2: Stationary Phase Particle Chromatographic Polarity Spectrum

Silica has an active, hydrophilic [water-loving] surface containing acidic silanol [silicon-containing analog of alcohol] functional groups. Consequently, it falls at the polar end of the stationary-phase scale shown in Figure R–2. The activity or polarity of the silica surface may be modified selectively by chemically bonding to it less polar functional groups [bonded phase]. Examples shown here include, in order of decreasing polarity, cyanopropylsilyl- [CN], n-octylsilyl- [C_8], and n-octadecylsilyl- [C_{18}, ODS] moieties on silica. The latter is a hydrophobic [water-hating], very non-polar packing.

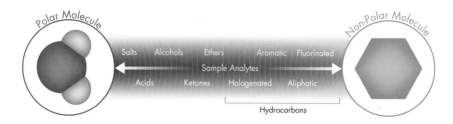

Figure R–3: Compound/Analyte Chromatographic Polarity Spectrum

Figure R–3 repeats the chromatographic polarity spectrum of our sample [shown in Figure P]. After considering the polarity of both phases, then, for a given stationary phase, a chromatographer must choose a mobile phase in which the analytes of interest are retained, but not so strongly that they cannot be eluted. Among solvents of similar strength, the chromatographer considers which phase combination may best exploit the more subtle differences in analyte polarity and solubility to maximize the selectivity of the chromatographic system. Like attracts like, but, as you probably can imagine from the discussion so far, creating a separation based upon polarity involves knowledge of the sample and experience with various kinds of analytes and retention modes. To summarize, the chromatographer will choose the best combination of a mobile phase and particle stationary phase with appropriately opposite polarities. Then, as the sample analytes move through the column, the rule *like attracts like* will determine which analytes slow down and which proceed at a faster speed.

Normal-Phase HPLC

In his separations of plant extracts, Tswett was successful using a polar stationary phase [chalk in a glass column; see Figure A] with a much less polar [non-polar] mobile phase. This classical mode of chromatography became known as normal phase.

Stationary Phase Is Polar [Silica]

Mobile Phase Is Non-Polar [Hexane]

Why Do They Separate?

Sample

Figure S–1: Normal-Phase Chromatography

Figure S–1 represents a normal-phase chromatographic separation of our three-dye test mixture. The stationary phase is polar and retains the polar yellow dye most strongly. The relatively non-polar blue dye is won in the retention competition by the mobile phase, a non-polar solvent, and elutes quickly. Since the blue dye is most like the mobile phase [both are non-polar], it moves faster. It is typical for normal-phase chromatography on silica that the mobile phase is 100% organic; no water is used.

Reversed-Phase HPLC

The term reversed-phase describes the chromatography mode that is just the opposite of normal phase, namely the use of a polar mobile phase and a non-polar [hydrophobic] stationary phase. Figure S–2 illustrates the black three-dye mixture being separated using such a protocol.

Stationary Phase Is Non-Polar [C_{18}]

Mobile Phase Is Polar [Aqueous]

Why Do They Separate?

Sample

Figure S–2: Reversed–Phase Chromatography

Now the most strongly retained compound is the more non-polar blue dye, as its attraction to the non-polar stationary phase is greatest. The polar yellow dye, being weakly retained, is won in competition by the polar, aqueous mobile phase, moves the fastest through the bed, and elutes earliest—*like attracts like*.

Today, because it is more reproducible and has broad applicability, reversed-phase chromatography is used for approximately 75% of all HPLC methods. Most of these protocols use as the mobile phase an aqueous blend of water with a miscible, polar organic solvent, such as acetonitrile or methanol. This typically ensures the proper interaction of analytes with the non-polar, hydrophobic particle surface. A C_{18}–bonded silica [sometimes called ODS] is the most popular type of reversed-phase HPLC packing.

Table C presents a summary of the phase characteristics for the two principal HPLC separation modes based upon polarity. Remember, for these polarity-based modes, *like attracts like*.

Separation Mode	Stationary Phase [particle]	Mobile Phase [solvent]
Normal phase	Polar	Non-polar
Reversed phase	Non-polar	Polar

Table C: Phase Characteristics for Separations Based on Polarity

Hydrophilic-Interaction Chromatography [HILIC]

HILIC may be viewed as a variant of normal-phase chromatography. It is used when polar analytes cannot be retained on a reversed-phased (non-polar) stationary phase.

In normal-phase chromatography, the mobile phase is 100% organic. Only traces of water are present in the mobile phase and in the pores of the polar packing particles. Polar analytes bind strongly to the polar stationary phase and may not elute.

Adding some water [< 20%] to the organic mobile phase [typically an aprotic solvent like acetonitrile] makes it possible to separate and elute polar compounds that are strongly retained in the normal-phase mode [or weakly retained in the reversed-phase mode]. Water, a very polar solvent, competes effectively with polar analytes for the stationary phase. HILIC may be run in either isocratic or gradient-elution modes. Polar compounds, that are initially attracted to the polar packing material particles, can be eluted as the polarity [strength] of the mobile phase is increased [by adding more water]. Analytes are eluted in order of increasing hydrophilicity [chromatographic polarity relative to water]. Buffers or salts may be added to the mobile phase to keep ionizable analytes in a single form.

Hydrophobic-Interaction Chromatography [HIC]

HIC is a type of reversed-phase chromatography that is used to separate large biomolecules, such as proteins. It is usually desirable to maintain these molecules intact in an aqueous solution, avoiding contact with organic solvents or surfaces that might denature them. HIC takes advantage of the hydrophobic interaction of large molecules with a moderately hydrophobic stationary phase, *e.g.*, butyl-bonded [C_4], rather than octadecyl-bonded [C_{18}], silica. Initially, higher salt concentrations in water will encourage the proteins to be retained [*salted out*] on the packing. Gradient separations are typically run by decreasing salt concentration. In this way, biomolecules are eluted in order of increasing hydrophobicity.

Separations Based on Charge: Ion-Exchange Chromatography [IEC]

For separations based on polarity, *like* is attracted to *like* and *opposites* may be repelled. In ion-exchange chromatography and other separations based upon electrical charge, the rule is reversed. *Likes may repel, while opposites are attracted to each other.* Stationary phases for ion-exchange separations are characterized by the nature and strength of the acidic or basic functions on their surfaces and the types of ions that they attract and retain. *Cation* exchange is used to retain and separate positively charged ions on a *negative* surface. Conversely, *anion* exchange is used to retain and separate negatively charged ions on a *positive* surface [see Figure T]. With each type of ion exchange, there are at least two general approaches for separation and elution.

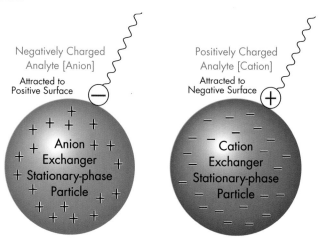

Figure T: Ion-Exchange Chromatography

Strong ion exchangers bear functional groups [*e.g.*, quaternary amines or sulfonic acids] that are always ionized. They are typically used to retain and separate *weak* ions. These weak ions may be eluted by displacement with a mobile phase containing ions that are more strongly attracted to the stationary phase sites. Alternately, weak ions may be retained on the column, then *neutralized* by *in situ* changing the pH of the mobile phase, causing them to lose their attraction and elute.

Weak ion exchangers [e.g., with secondary-amine or carboxylic-acid functions] may be neutralized above or below a certain pH value and lose their ability to retain ions by charge. When charged, they are used to retain and separate strong ions. If these ions cannot be eluted by displacement, then the stationary phase exchange sites may be neutralized, shutting off the ionic attraction, and permitting elution of the charged analytes.

Analyte Type	Weak ACID e.g., pKa ~ 5		Strong ACID	Weak BASE e.g., pKa ~ 10		Strong BASE
Charge State vs. pH*	No charge at pH < 3	— [anion] at pH > 7	— [anion] Always Charged	+ [cation] at pH < 8	No Charge at pH > 12	+ [cation] Always Charged

Stationary Phase Particle	Strong Anion Exchanger	Weak Anion Exchanger e.g., pKa ~ 10		Strong Cation Exchanger	Weak Cation Exchanger e.g., pKa ~ 5	
Charge State vs. pH*	+ Always Charged	+ at pH < 8	No Charge at pH > 12	— Always Charged	No Charge at pH < 3	— at pH > 7

Mobile Phase pH Range				
to Retain analyte [capture]	pH > 7	pH < 8	pH < 8	pH > 7
to Release analyte [elute]	pH < 3	pH > 12	pH > 12	pH < 3

*Note: pH Ranges are approximate. They will depend upon specific analyte and particle characteristics.

Table D: Ion-Exchange Guidelines

First, determine analyte type. Then, follow corresponding arrows down for recommended particle and mobile phase pH.

When weak ion exchangers are *neutralized*, they may retain and separate species by *hydrophobic* [reversed-phase] or *hydrophilic* [normal-phase] interactions; in these cases, elution strength is determined by the polarity of the mobile phase [Figure R−1]. Thus, weak ion exchangers may be used for mixed-mode separations [separations based on both polarity and charge].

Table D outlines guidelines for the principal categories of ion exchange. For example, to retain a *strongly basic* analyte [always positively charged], use a *weak-cation*-exchange stationary phase particle at pH > 7; this assures a *negatively* charged particle surface. To release or elute the strong base, lower the pH of the mobile phase below 3; this removes the surface charge and *shuts off* the ion-exchange retention mechanism.

Note that a pK_a is the pH value at which 50% of the functional group is ionized and 50% is neutral. To assure an essentially neutral, or a fully charged, analyte or particle surface, the pH must be adjusted to a value at least 2 units beyond the pK_a, as appropriate [indicated in Table D].

Do not use a strong-cation exchanger to retain a strong base; both remain charged and strongly attracted to each other, making the base nearly impossible to elute. It can only be removed by swamping the strong cation exchanger with a competing base that exhibits even stronger retention and displaces the compound of interest by winning the competition for the active exchange sites. This approach is rarely practical, or safe, in HPLC and SPE. [Very strong acids and bases are dangerous to work with, and they may be corrosive to materials of construction used in HPLC fluidics!]

Separations Based on Size: Size-Exclusion Chromatography [SEC] – Gel-Permeation Chromatography [GPC]

In the 1950s, Porath and Flodin discovered that biomolecules could be separated based on their size, rather than on their charge or polarity, by passing, or *filtering*, them through a controlled-porosity, hydrophilic dextran polymer. This process was termed *gel filtration*. Later, an analogous scheme was used to separate synthetic oligomers and polymers using organic-polymer packings with specific pore-size ranges. This process was called gel-permeation chromatography [GPC]. Similar separations done using controlled-porosity silica packings were called size-exclusion chromatography [SEC]. Introduced in 1963, the first commercial HPLC instruments were designed for GPC applications [see Reference 3].

All of these techniques are typically done on stationary phases that have been synthesized with a pore-size distribution over a range that permits the analytes of interest to enter, or to be excluded from, more or less of the pore volume of the packing. Smaller molecules penetrate more of the pores on their passage through the bed. Larger molecules may only penetrate pores above a certain size so they spend less time in the bed. The biggest molecules may be totally excluded from pores and pass only between the particles, eluting very quickly in a small volume. Mobile phases are chosen for two reasons: first, they are good solvents for the analytes; and, second, they may prevent any interactions [based on polarity or charge] between the analytes and the stationary phase surface. In this way, the larger molecules elute first, while the smaller molecules travel slower [because they move into and out of more of the pores] and elute later, in decreasing order of their size in solution. Hence the simple rule: *Big ones come out first*.

Since it is possible to correlate the molecular weight of a polymer with its size in solution, GPC revolutionized measurement of the molecular-weight distribution of polymers, that, in turn, determines the physical characteristics that may enhance, or detract from, polymer processing, quality, and performance [how to tell *good* from *bad* polymer].

Conclusion

We hope you have enjoyed this brief introduction to HPLC. We encourage you to read the references that follow and to study the Appendix on HPLC Nomenclature. You may also access over 50,000 references to applications of HPLC, SPE, and MS in the Waters Library at <http://www.waters.com>.

References for Further Reading:

1. U.D. Neue, "HPLC Columns: Theory, Technology, and Practice," Wiley-VCH [1997]

2. P.D. McDonald and B.A. Bidlingmeyer, "Strategies for Successful Preparative Liquid Chromatography," Chap. 1 in: *J. Chromatogr. Lib.* **38**: 1–103 [1987]

3. P.D. McDonald, "James Waters and His Liquid Chromatography People: a Personal Perspective," *Waters Whitepaper* **WA62008***: 20 pp [2006]

   *Use this code to access the PDF document in Waters Library at <http://www.waters.com>

4. P.D. McDonald, "Improving Our Understanding of Reversed-Phase Separations for the 21st Century," Chap. 7 in: *Adv. Chromatogr.* **42**: 323–375 [2003]

5. J.C. Arsenault, E.S. Grumbach, D.R. McCabe, and P.D. McDonald, "Achieving Ultra Performance in Liquid Chromatography: An Introduction to UPLC Technology," Waters, in press [2009]

6. M.P. Balogh, "The Mass Spectrometry Primer," Waters [2009]

Appendix: HPLC Nomenclature

*Indicates a definition adapted from: L.S. Ettre, *Nomenclature for Chromatography, Pure Appl. Chem.* **65:** 819-872 [1993], © 1993 IUPAC; an updated version of this comprehensive report is available in the *Orange Book*, Chapter 9: *Separations* [1997] at: <http://www.iupac.org/publications/analytical_compendium>.

Alumina

A porous, particulate form of aluminum oxide [Al_2O_3] used as a stationary phase in normal-phase adsorption chromatography. Alumina has a highly active basic surface; the pH of a 10% aqueous slurry is about 10. It is successively washed with strong acid to make neutral and acidic grades [slurry pH 7.5 and 4, resp.]. Alumina is more hygroscopic than silica. Its activity is measured according to the Brockmann[†] scale for water content; *e.g.,* Activity Grade I contains 1% H_2O.

[†]H. Brockmann and H. Schodder, Ber. 74: 73 (1941).

Baseline*

The portion of the chromatogram recording the detector response when only the mobile phase emerges from the column.

Cartridge

A type of column, without endfittings, that consists simply of an open tube wherein the packing material is retained by a frit at either end. SPE cartridges may be operated in parallel on a vacuum-manifold. HPLC cartridges are placed into a cartridge holder that has fluid connections built into each end. Cartridge columns are easy to change, less expensive, and more convenient than conventional columns with integral endfittings.

Chromatogram*

A graphical or other presentation of detector response or other quantity used as a measure of the concentration of the analyte in the effluent versus effluent volume or time. In planar chromatography [*e.g.,* thin-layer chromatography or paper chromatography], *chromatogram* may refer to the paper or layer containing the separated zones.

Chromatography*

A dynamic physicochemical method of separation in which the components to be separated are distributed between two phases, one of which is stationary [the *stationary phase*] while the other [the *mobile phase*] moves relative to the stationary phase.

Column Volume* [V_C]

The geometric volume of the part of the tube that contains the packing [internal cross-sectional area of the tube multiplied by the packed bed length, L]. The *interparticle volume* of the column, also called the *interstitial volume*, is the volume occupied by the mobile phase between the particles in the packed bed. The void volume [V_0] is the total volume occupied by the mobile phase, *i.e.* the sum of the interstitial volume and the *intraparticle volume* [also called *pore volume*].

Detector* [see Sensitivity]

A device that indicates a change in the composition of the eluent by measuring physical or chemical properties [*e.g.,* UV/visible light absorbance, differential refractive index, fluorescence, or conductivity]. If the detector's response is linear with respect to sample concentration, then, by suitable calibration with standards, the amount of a component may be quantitated. Often, it may be beneficial to use two differ-ent types of detectors in series. In this way, more corroboratory or specific information may be obtained about the sample analytes. Some detectors [*e.g.,* electrochemical, mass spectrometric] are destructive; *i.e.*, they effect a chemical change in the sample components. If a detector of this type is paired with a non-destructive detector, it is usually placed second in the flow path.

Display

A device that records the electrical response of a detector on a computer screen in the form of a chro-matogram. Advanced data recording systems also perform calculations using sophisticated algorithms, *e.g.,* to integrate peak areas, subtract baselines, match spectra, quantitate components, and identify unknowns by comparison to standard libraries.

Efficiency [H, see Plate Number, Resolution, Sensitivity, Speed]

A measure of a column's ability to resist the dispersion of a sample band as it passes through the packed bed. An efficient column minimizes *band dispersion* or *bandspreading*. Higher efficiency is important for effective separation, greater sensitivity, and/or identification of similar components in a complex sample mixture.

Nobelists Martin and Synge, by analogy to distillation, introduced the concept of *plate height* [H, or H.E.T.P., *height equivalent to a theoretical plate*] as a measure of chromatographic efficiency and as a means to compare column performance.[†] Presaging HPLC and UPLC® technology, they recognized that a homogeneous bed packed with the smallest possible particle size [requiring higher pressure] was key to maximum efficiency. The relation between column and separation system parameters that affect band-spreading was later described in an equation by van Deemter.[††]

Chromatographers often refer to a quantity that they can calculate easily and directly from measurements made on a chromatogram, namely *plate number* [N], as efficiency. Plate height is then determined from the ratio of the length of the column bed to N [H = L/N; methods of calculating N from a chromatogram are shown in Figure U]. It is important to note that calculation of N or H using these methods is correct only for isocratic conditions and cannot be used for gradient separations.

[†]A.J.P. Martin and R.M. Synge, *Biochem. J.* **35**: 1358-1368 [1941]

[††]J.J. van Deemter, F. J. Zuiderweg and A. Klinkenberg, *Chem. Eng. Sci.* **5**: 271-289 [1956]

Eluate

The portion of the *eluent* that emerges from the column outlet containing analytes in solution. In analytical HPLC, the *eluate* is examined by the detector for the concentration or mass of analytes therein. In preparative HPLC, the eluate is collected continuously in aliquots at uniform time or volume intervals, or discontinuously only when a detector indicates the presence of a peak of interest. These fractions are subsequently processed to obtain purified compounds.

Eluent

The *mobile phase* [see Elution Chromatography].

Eluotropic Series

A list of solvents ordered by *elution strength* with reference to specified analytes on a standard sorbent. Such a series is useful when developing both isocratic and gradient elution methods. Trappe coined this term after showing that a sequence of solvents of increasing polarity could separate lipid fractions on alumina.[†] Later, Snyder measured and tabulated solvent strength parameters for a large list of solvents on several normal-phase LC sorbents.[††] Neher created a very useful *nomogram* by which *equi-eluotropic* [constant elution strength] mixtures of normal-phase solvents could be chosen to optimize the selectivity of TLC separations.[†††]

A typical normal-phase *eluotropic series* would start at the weak end with non-polar aliphatic hydrocarbons, *e.g.,* pentane or hexane, then progress successively to benzene [an aromatic hydrocarbon], dichloromethane [a chlorinated hydrocarbon], diethyl ether, ethyl acetate [an ester], acetone [a ketone], and, finally, methanol [an alcohol] at the strong end [see Figure R-1].

[†]W. Trappe, *Biochem. Z.* **305**: 150 [1940]

[††]L. R. Snyder, *Principles of Adsorption Chromatography,* Marcel Dekker [1968], pp. 192-197

[†††]R. Neher in G.B. Marini-Bettòlo, ed., *Thin-Layer Chromatography,* Elsevier [1964] pp. 75-86.

Elute* [verb]

To chromatograph by elution chromatography. The process of elution may be stopped while all the sample components are still on the chromatographic bed [planar thin-layer or paper chromatography] or continued until the components have left the chromatographic bed [column chromatography].

Note: The term *elute* is preferred to *develop* [a term used in planar chromatography], to avoid confusion with the practice of method development, whereby a separation system [the combination of mobile and stationary phases] is optimized for a particular separation.

Elution Chromatography*

A procedure for chromatographic separation in which the mobile phase is continuously passed through the chromatographic bed. In HPLC, once the detector baseline has stabilized and the separation system has reached equilibrium, a finite slug of sample is introduced into the flowing mobile phase stream. Elution continues until all analytes of interest have passed through the detector.

Elution Strength

A measure of the affinity of a solvent relative to that of the analyte for the stationary phase. A weak solvent cannot displace the analyte, causing it to be strongly retained on the stationary phase. A strong solvent may totally displace all the analyte molecules and carry them through the column unretained. To achieve a proper balance of effective separation and reasonable elution volume, solvents are often blended to set up an appropriate *competition* between the phases, thereby optimizing both selectivity and separation time for a given set of analytes [see Selectivity].

Dipole moment, dielectric constant, hydrogen bonding, molecular size and shape, and surface tension may give some indication of elution strength. Elution strength is also determined by the separation mode. An *eluotropic series* of solvents may be ordered by increasing strength in one direction under *adsorption* or *normal-phase* conditions; that order may be nearly opposite under *reversed-phase partition* conditions [see Figure R-1].

Fluorescence Detector

Fluorescence detectors *excite* a sample with a specified wavelength of light. This causes certain compounds to fluoresce and emit light at a higher wavelength. A sensor, set to a specific *emission wavelength* and masked so as not to be blinded by the excitation source, collects only the emitted light. Often analytes that do not natively fluoresce may be derivatized to take advantage of the high sensitivity and selectivity of this form of detection, *e.g.,* AccQ•Tag™ derivatization of amino acids.

Flow Rate*

The volume of mobile phase passing through the column in unit time. In HPLC systems, the flow rate is set by the controller for the solvent delivery system [pump]. Flow rate accuracy can be checked by timed collection and measurement of the effluent at the column outlet. Since a solvent's density varies with temperature, any calibration or flow rate measurement must take this variable into account. Most accurate determinations are made, when possible, by weight, not volume.

Uniformity [precision] and *reproducibility* of flow rate is important to many LC techniques, especially in separations where *retention times* are key to analyte identification, or in *gel-permeation chromatography* where calibration and correlation of retention times are critical to accurate molecular-weight-distribution measurements of polymers.

Often, separation conditions are compared by means of *linear velocity*, not flow rate. The linear velocity is calculated by dividing the flow rate by the cross-sectional area of the column. While flow rate is expressed in volume/time [*e.g.*, mL/min], linear velocity is measured in length/time [*e.g.*, mm/sec].

Gel-Permeation Chromatography*

Separation based mainly upon exclusion effects due to differences in molecular size and/or shape. *Gel-permeation chromatography* and *gel filtration chromatography* describe the process when the stationary phase is a swollen gel. Both are forms of *size-exclusion chromatography*. Porath and Flodin first described gel-filtration using dextran gels and aqueous mobile phases for the size-based separation of biomolecules.[†] Moore applied similar principles to the separation of organic polymers by size in solution using organic-solvent mobile phases on porous polystyrene-divinylbenzene polymer gels.[††]

[†] J. Porath, P. Flodin, *Nature* **183**: 1657-1659 [1959]

[††] J.C. Moore, *U.S. Patent* 3,326,875 [filed Jan. 1963; issued June 1967]

Gradient

The change over time in the relative concentrations of two [or more] miscible solvent components that form a mobile phase of increasing elution strength. A *step gradient* is typically used in solid-phase extraction; in each step, the eluent composition is changed abruptly from a weaker mobile phase to a stronger mobile phase. It is even possible, by drying the SPE sorbent bed in between steps, to change from one solvent to another immiscible solvent.

A *continuous* gradient is typically generated by a low- or high-pressure mixing system [see Figures J-2 and J-3] according to a pre-determined curve [linear or non-linear] representing the concentration of the stronger solvent B in the initial solvent A over a fixed time period. A *hold* at a fixed *isocratic* solvent composition can be programmed at any time point within a continuous gradient. At the end of a separation, the *gradient program* can also be set to return to the initial mobile phase composition to re-equilibrate the column in preparation for the injection of the next sample. Sophisticated HPLC systems can blend as many as four or more solvents [or solvent mixtures] into a continuous gradient.

Injector [Autosampler, Sample Manager]

A mechanism for accurately and precisely introducing [*injecting*] a discrete, predetermined volume of a sample solution into the flowing mobile phase stream. The injector can be a simple manual device, or a sophisticated autosampler that can be programmed for unattended injections of many samples from an array of individual vials or wells in a predetermined sequence. Sample compartments in these systems may even be temperature controlled to maintain sample integrity over many hours of operation.

Most modern injectors incorporate some form of syringe-filled sample loop that can be switched on- or offline by means of a multi-port valve. A well-designed, minimal-internal-volume injection system is situated as close to the column inlet as possible and minimizes the spreading of the sample band. Between sample injections, it is also capable of being flushed to waste by mobile phase, or a wash solvent, to prevent *carryover* [contamination of the present sample by a previous one].

Samples are best prepared for injection, if possible, by dissolving them in the mobile phase into which they will be injected; this may prevent issues with separation and/or detection. If another solvent must be used, it is desirable that its elution strength be equal to or less than that of the mobile phase. It is often wise to mix a bit of a sample solution with the mobile phase offline to test for precipitation or miscibility issues that might compromise a successful separation.

Inlet

The end of the column bed where the mobile phase stream and sample enter. A porous, inert frit retains the packing material and protects the sorbent bed inlet from particulate contamination. Good HPLC practice dictates that samples and mobile phases should be particulate-free; this becomes imperative for small-particle columns whose inlets are much more easily plugged. If the column bed inlet becomes clogged and exhibits higher-than-normal *backpressure*, sometimes, reversing the flow direction while directing the effluent to waste may dislodge and flush out sample debris that sits atop the frit. If the debris has penetrated the frit and is lodged in the inlet end of the bed itself, then the column has most likely reached the end of its useful life.

Ion-Exchange Chromatography* [see section: Separations Based on Charge]

This separation mode is based mainly on differences in the ion-exchange affinities of the sample components.

Separation of primarily inorganic ionic species in water or buffered aqueous mobile phases on small-particle, superficially porous, high-efficiency, ion-exchange columns followed by conductometric or electrochemical detection is referred to as ion chromatography [IC].

Isocratic Elution*

A procedure in which the composition of the mobile phase remains constant during the elution process.

Liquid Chromatography* [LC]

A separation technique in which the mobile phase is a liquid. Liquid chromatography can be carried out either in a column or on a plane [TLC or paper chromatography]. Modern liquid chromatography utilizing smaller particles and higher inlet pressure was termed *high-performance* (or *high-pressure*) *liquid chromatography* [HPLC] in 1970. In 2004, *ultra-performance liquid chromatography* dramatically raised the performance of LC to a new plateau [see UPLC® Technology].

Mobile Phase* [see Eluate, Eluent]

A fluid that percolates, in a definite direction, through the length of the stationary-phase sorbent bed. The mobile phase may be a liquid [*liquid chromatography*] or a gas [*gas chromatography*] or a supercritical fluid [*supercritical-fluid chromatography*]. In gas chromatography the expression *carrier gas* may be used for the mobile phase. In elution chromatography, the mobile phase may also be called the *eluent*, while the word *eluate* is defined as the portion of the mobile phase that has passed through the sorbent bed and contains the compounds of interest in solution.

Normal-Phase Chromatography*

An elution procedure in which the stationary phase is more polar than the mobile phase. This term is used in liquid chromatography to emphasize the contrast to *reversed-phase chromatography*.

Peak* [see Plate Number]

The portion of a differential chromatogram recording the detector response while a single component is eluted from the column. If separation is incomplete, two or more components may be eluted as one *unresolved* peak. Peaks eluted under optimal conditions from a well-packed, efficient column, operated in a system that minimizes bandspreading, approach the shape of a Gaussian distribution. Quantitation is usually done by measuring the *peak area* [enclosed by the baseline and the peak curve]. Less often, peak height [the distance measured from the peak apex to the baseline] may be used for quantitation. This procedure requires that both the peak width and the peak shape remain constant.

Plate Number* [N, see Efficiency]

A number indicative of column performance [mechanical separation power or efficiency, also called *plate count, number of theoretical plates,* or *theoretical plate number*]. It relates the magnitude of a peak's retention to its width [*variance* or *bandspread*]. In order to calculate a plate count, it is assumed that a peak can be represented by a Gaussian distribution [a statistical *bell curve*]. At the inflection points [60.7% of peak height], the width of a Gaussian curve is twice the *standard deviation* [σ] about its mean [located at the peak apex]. As shown in Figure U, a Gaussian curve's peak width measured at other fractions of peak height can be expressed in precisely defined multiples of σ. Peak retention [retention volume, V_R, or retention time, t_R] and peak width must be expressed in the same units, because N is a dimensionless number. Note that the 5σ method of calculating N is a more stringent measure of column homogeneity and performance, as it is more severely affected by peak asymmetry. Computer data stations can automatically delineate each resolved peak and calculate its corresponding plate number.

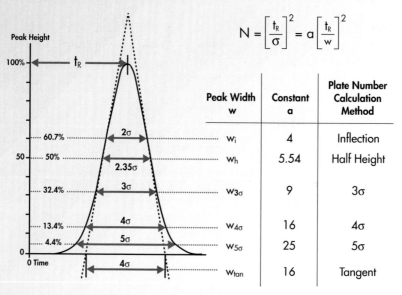

$$N = \left[\frac{t_R}{\sigma}\right]^2 = a\left[\frac{t_R}{w}\right]^2$$

Peak Width w	Constant a	Plate Number Calculation Method
w_i	4	Inflection
w_h	5.54	Half Height
$w_{3\sigma}$	9	3σ
$w_{4\sigma}$	16	4σ
$w_{5\sigma}$	25	5σ
w_{tan}	16	Tangent

Figure U: Methods for Calculating Plate Number [N]

Preparative Chromatography

The process of using liquid chromatography to isolate a compound in a quantity and at a purity level sufficient for further experiments or uses. For pharmaceutical or biotechnological purification processes, columns several feet in diameter can be used for multiple kilograms of material. For isolating just a few micrograms of a valuable natural product, an analytical HPLC column is sufficient. Both are preparative chromatographic approaches, differing only in scale [see section on HPLC Scale and Table A].

Resolution* [R_s, see Selectivity]

The separation of two peaks, expressed as the difference in their corresponding retention times, divided by their average peak width at the baseline. $R_s = 1.25$ indicates that two peaks of equal width are just separated at the baseline. When $R_s = 0.6$, the only visual indication of the presence of two peaks on a chromatogram is a small notch near the peak apex. Higher efficiency columns produce narrower peaks and improve resolution for difficult separations; however, *resolution* increases by only the *square root of N*. The most powerful method of increasing resolution is to increase *selectivity* by altering the mobile/stationary phase combination used for the chromatographic separation [see section on Chemical Separation Power].

Retention Factor* [k]

A measure of the time the sample component resides in the stationary phase relative to the time it resides in the mobile phase; it expresses how much longer a sample component is retarded by the stationary phase than it would take to travel through the column with the velocity of the mobile phase. Mathematically, it is the ratio of the adjusted retention time [volume] and the hold-up time [volume]: $k = t_R'/t_M$ [see Retention Time and Selectivity].

Note: In the past, this term has also been expressed as p*artition ratio*, *capacity ratio*, *capacity factor*, or *mass distribution ratio* and symbolized by k'.

Retention Time* [t_R]

The time between the start of elution [typically, in HPLC, the moment of injection or sample introduction] and the emergence of the peak maximum. The *adjusted retention time*, t_R', is calculated by subtracting from t_R the *hold-up time* [t_M, the time from injection to the elution of the peak maximum of a totally unretained analyte].

Reversed-Phase Chromatography*

An elution procedure used in liquid chromatography in which the mobile phase is significantly more polar than the stationary phase, *e.g.* a microporous silica-based material with alkyl chains chemically bonded to its accessible surface. Note: Avoid the incorrect term *reverse phase*. [See Reference 4 for some novel ideas on the mechanism of reversed-phase separations.]

Selectivity [Separation Factor, α]

A term used to describe the magnitude of the difference between the relative thermodynamic affinities of a pair of analytes for the specified mobile and stationary phases that comprise the separation system. The proper term is *separation factor* [α]. It equals the ratio of retention factors, k_2/k_1 [see Retention Factor]; by definition, α is always ≥ 1. If $\alpha = 1$, then both peaks co-elute, and no separation is obtained. It is important in preparative chromatography to maximize α for highest sample loadability and throughput. [see section on Chemical Separation Power]

Sensitivity* [S]

The signal output per unit concentration or unit mass of a substance in the mobile phase entering the detector, *e.g.,* the slope of a linear calibration curve [see Detector]. For concentration-sensitive detectors [*e.g.,* UV/VIS absorbance], sensitivity is the ratio of peak height to analyte concentration in the peak. For mass-flow-sensitive detectors, it is the ratio of peak height to unit mass. If sensitivity is to be a unique performance characteristic, it must depend only on the chemical measurement process, not upon scale factors.

The ability to detect [*qualify*] or measure [*quantify*] an analyte is governed by many instrumental and chemical factors. Well-resolved peaks [maximum selectivity] eluting from high efficiency columns [narrow peak width with good symmetry for maximum peak height] as well as good detector sensitivity and specificity are ideal. Both the separation system interference and electronic component noise should also be minimized to achieve maximum sensitivity.

Solid-Phase Extraction [SPE]

A sample preparation technique that uses LC principles to isolate, enrich, and/or purify analytes from a complex matrix applied to a miniature chromatographic bed. *Offline* SPE is done [manually or *via* automation] with larger particles in individual plastic cartridges or in micro-elution plate wells, using low positive pressure or vacuum to assist flow. *Online* SPE is done with smaller particles in miniature HPLC columns using higher pressures and a valve to switch the SPE column online with the primary HPLC column, or offline to waste, as appropriate.

SPE methods use step gradients [see Gradient] to accomplish bed conditioning, sample loading, washing, and elution steps. Samples are loaded typically under conditions where the *k* of important analytes is as high as possible, so that they are fully retained during loading and washing steps. Elution is then done by switching to a much stronger solvent mixture [see Elution Strength]. The goal is to remove matrix interferences and to isolate the analyte in a solution, and at a concentration, suitable for subsequent analysis.

Speed [see Efficiency, Flow Rate, Resolution]

A benefit of operating LC separations at higher *linear velocities* using smaller-volume, smaller-particle analytical columns, or larger-volume, larger-particle preparative columns. Order-of-magnitude advances in LC speed came in 1972 [with the use of 10 µm particles and pumps capable of delivering accurate mobile-phase flow at 6000 psi], in 1976 [with 75-µm preparative columns operated at a flow rate of 500 mL/min], and in 2004 [with the introduction of UPLC® technology—1.7 µm-particle columns operated at 15,000 psi].[†]

High-speed analytical LC systems must not only accommodate higher pressures throughout the fluidics; injector cycle time must be short; gradient mixers must be capable of rapid turnaround between samples; detector sensors must rapidly respond to tiny changes in eluate composition; and data systems must collect the dozens of points each second required to plot and to quantitate narrow peaks accurately.

Together, higher resolution, higher speed, and higher efficiency typically deliver higher *throughput*. More samples can be analyzed in a workday. Larger quantities of compound can be purified per run or per process period.

[†]See #3 on list of *References for Further Reading* above.

Stationary Phase[*]

One of the two phases forming a chromatographic system. It may be a solid, a gel, or a liquid. If a liquid, it may be distributed on a solid. This solid may or may not contribute to the separation process. The liquid may also be chemically bonded to the solid [*bonded phase*] or immobilized onto it [*immobilized phase*].

The expression *chromatographic bed* or *sorbent* may be used as a general term to denote any of the different forms in which the stationary phase is used.

The use of the term *liquid phase* to denote the *mobile phase* in LC is discouraged. This avoids confusion with gas chromatography where the *stationary phase* is called a *liquid phase* [most often a liquid coated on a solid support].

Open-column liquid-liquid partition chromatography [LLC] did not translate well to HPLC. It was supplanted by the use of bonded-phase packings. LLC proved incompatible with modern detectors because of problems with bleed of the stationary-phase-liquid coating off its solid support, thereby contaminating the immiscible liquid mobile phase.

UPLC® Technology

The use of a high-efficiency LC system holistically designed to accommodate sub-2 μm particles and very high operating pressure is termed *ultra-performance liquid chromatography* [UPLC® technology].[†] The major benefits of this technology are significant improvements in resolution over HPLC, and/or faster run times while maintaining the resolution seen in an existing HPLC separation.

[†]For more information, visit: <www.waters.com/uplc> Also, see #5 on list of *References for Further Reading* above.